IXL MATH WORKBOOK

KINDERGARTEN ADDITION

ISBN: 9781947569539
29 28 27 26 25 4 5 6 7 8

Printed in China

2 Counting to 10

Time to warm up! Count the objects. Write the number.

4

IXL.com
skill ID
WVJ

For more practice, visit IXL.com or the IXL mobile app and enter this code in the search bar.

Shade the number of objects.

3

1

4

9

7

8

IXL.com skill ID **VNE**

4 Modeling addition

Write the total.

1 and 3 is 4

2 and 3 is ____

2 and 5 is ____

Write the total.

5 and 1 is ______

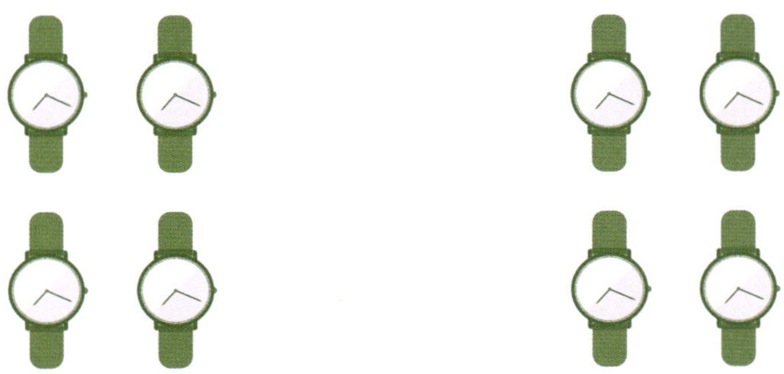

4 and 4 is ______

4 and 5 is ______

6 Modeling addition

Write each number to find the total.

2 and 4 is 6

____ and ____ is ____

____ and ____ is ____

Write each number to find the total.

____ and ____ is ____

____ and ____ is ____

8 Modeling addition

Write each number to find the total.

_____ and _____ is _____

_____ and _____ is _____

_____ and _____ is _____

Write each number to find the total.

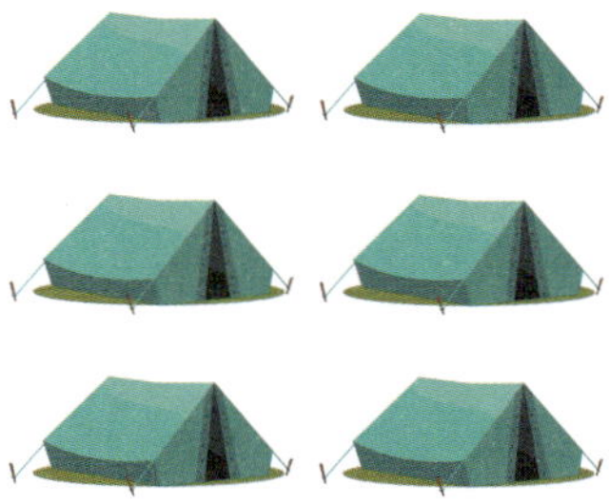

________ and ________ is ________

________ and ________ is ________

Addition sentences

Let's Learn!

You can use **+** to show groups coming together. It is called the **plus sign**.

2 + 3 is 5

plus

Write the numbers and the plus sign. Then write the total.

2 + 1 is 3

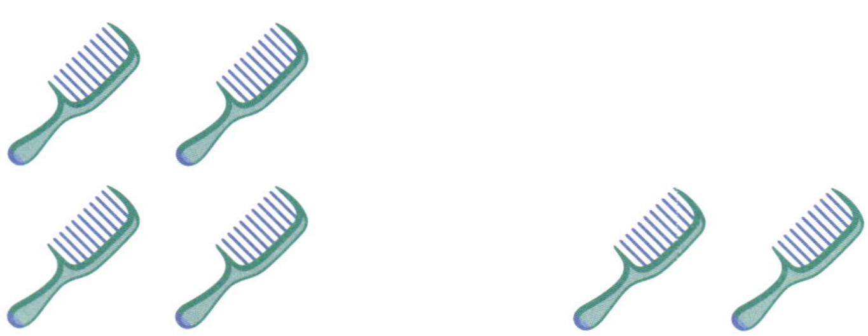

is

Write the numbers and the plus sign. Then write the total.

is

is

is

12 Addition sentences

Let's Learn!

You can use = to show that two things are the same. It is called the **equal sign**. You can use the **plus sign** and the **equal sign** to write an addition sentence.

1 + 3 = 4

plus equals

Write each addition sentence.

1 + 4 = 5

Write each addition sentence.

IXL.com
skill ID
KM7

Word problems

Write each addition sentence to solve.

There are 3 dogs at the park. Then 1 more dog joins them. How many dogs are at the park now?

3 + 1 = 4 dogs

Maggie planted 4 flowers. Then she planted 2 more flowers. How many flowers did Maggie plant?

________________ flowers

Write each addition sentence to solve.

There are 4 cows in a field. Then 4 more cows join them. How many cows are there in all?

____

______________________________ cows

____

Leo saw 2 small fish and 6 big fish. How many fish did he see in all?

____

______________________________ fish

____

IXL.com skill ID **7VF**

16 Sums of 4 and 5

Show some ways to make 4 and 5. Write each addition sentence.

3 + 2 = 5

Write the addition sentence for each animal.

4 + 1 = 5

Sums to 5

Draw to add.

$1 + 2 = 3$

$1 + 1 =$ ____

$4 + 1 =$ ____

Draw to add.

$2 + 2 =$ ____

$3 + 1 =$ ____

$2 + 3 =$ ____

IXL.com
skill ID
6RS

Add. Draw objects to help.

$1 + 4 = 5$

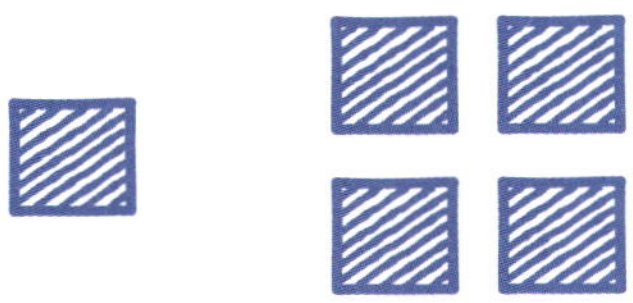

$1 + 1 =$ ______

$1 + 2 =$ ______

$3 + 1 =$ ______

$2 + 2 =$ ______

$2 + 3 =$ ______

Add.

$1 + 3 =$ ______ $4 + 1 =$ ______

$2 + 1 =$ ______ $1 + 1 =$ ______

$3 + 2 =$ ______ $2 + 2 =$ ______

$1 + 2 =$ ______ $2 + 3 =$ ______

$1 + 4 =$ ______ $3 + 1 =$ ______

22 Word problems

Write each addition sentence to solve. Draw a picture to help.

Jimmy ate 1 egg for breakfast. Adam ate 2 eggs for breakfast. How many eggs did Jimmy and Adam eat in all?

1 + 2 = 3 eggs

Rita put 2 candles on a birthday cake. Then she put 2 more candles on the cake. How many candles did Rita put on the cake?

candles

Write each addition sentence to solve. Draw a picture to help.

Zara and Andrew are painting. Zara puts 3 handprints on her painting. Andrew puts 1 handprint on his painting. How many handprints are there in all?

____________________ handprints

Molly saw 2 blue kites in the sky. Ryan saw 3 red kites. How many kites did Molly and Ryan see in all?

____________________ kites

IXL.com skill ID **QDH**

24 Sums of 6 and 7

Show some ways to make 6 and 7. Write each addition sentence.

3 + 4 = 7

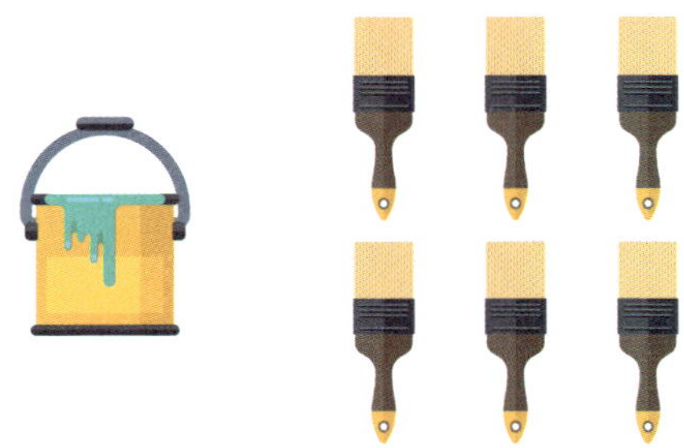

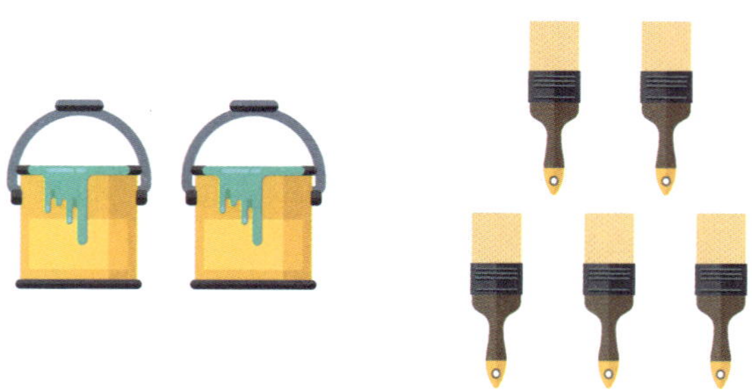

Show some ways to make 6 and 7. Write each addition sentence.

26 Sums of 6 and 7

Draw to add.

$4 + 3 = 7$

$3 + 3 =$

$5 + 1 =$

Draw to add.

$5 + 2 =$ ______

$2 + 4 =$ ______

$6 + 1 =$ ______

28 Sums of 6 and 7

Add.

$2 + 4 =$ ____

$4 + 3 =$ ____

$5 + 1 =$ ____

$5 + 2 =$ ____

$6 + 1 =$ ____

$3 + 3 =$ ____

$3 + 4 =$ ____

$4 + 2 =$ ____

$1 + 5 =$ ____

$2 + 5 =$ ____

Let's review! Add.

$1 + 2 =$ ______

$3 + 1 =$ ______

$3 + 4 =$ ______

$1 + 5 =$ ______

$2 + 2 =$ ______

$4 + 1 =$ ______

$3 + 2 =$ ______

$2 + 5 =$ ______

$1 + 6 =$ ______

$4 + 2 =$ ______

30 Word problems

Write each addition sentence to solve.

Scott baked 1 cake in the morning. He baked 3 more cakes in the afternoon. How many cakes did Scott bake?

1 + 3 = 4 cakes

There are 4 turtles swimming in a lake. Then 2 more turtles join them. How many turtles are swimming in the lake now?

__________ turtles

Cole found 5 pine cones on a walk. Then he found 2 more pine cones at the playground. How many pine cones did Cole find in total?

__________ pine cones

Write each addition sentence to solve.

In a bird's nest, there are 3 blue eggs. There are also 2 brown eggs. How many eggs are in the nest?

____________________ eggs

Maria has 4 red apples and 3 green apples. How many apples does Maria have?

____________________ apples

32 Sums of 8 and 9

Show some ways to make 8 and 9. Write each addition sentence.

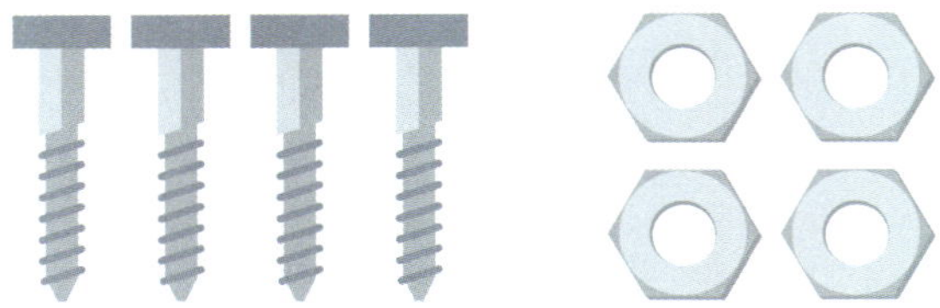

4 + 4 = 8

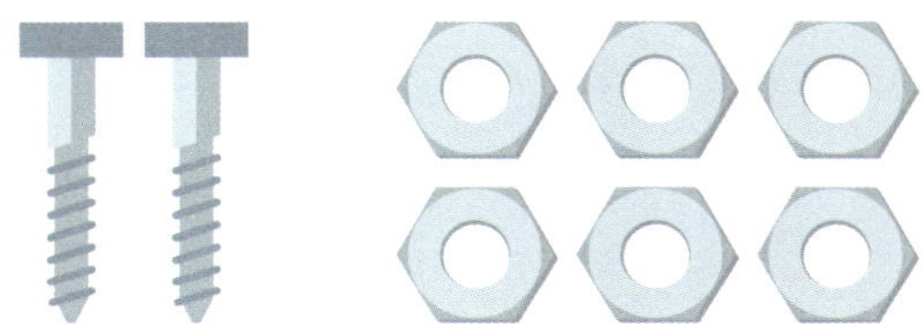

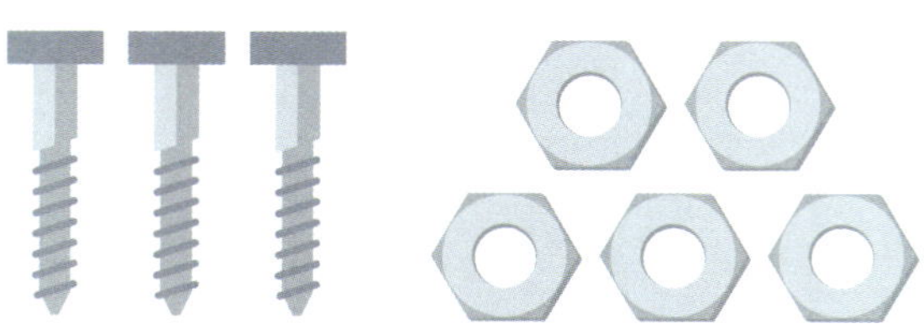

Show some ways to make 8 and 9. Write each addition sentence.

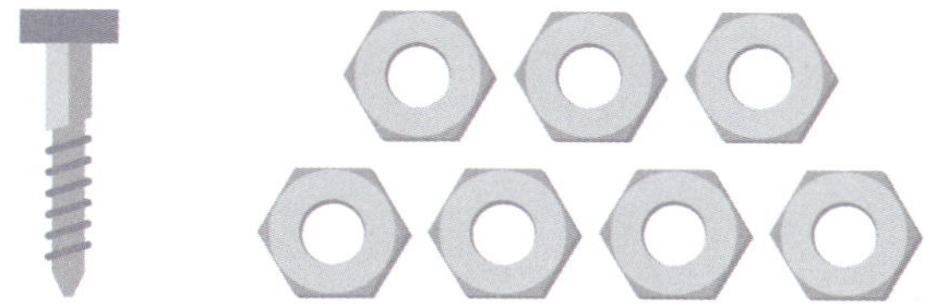

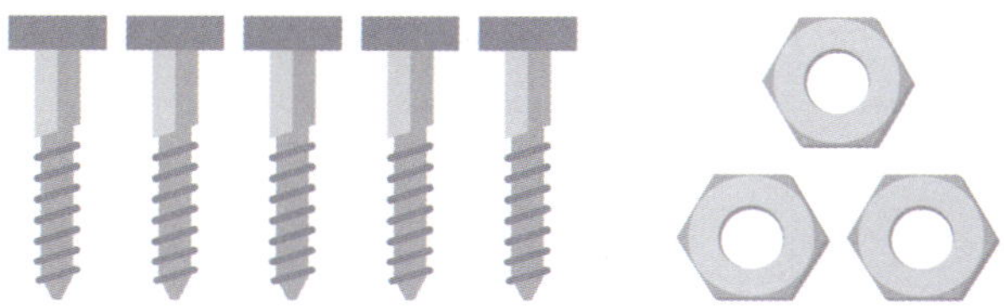

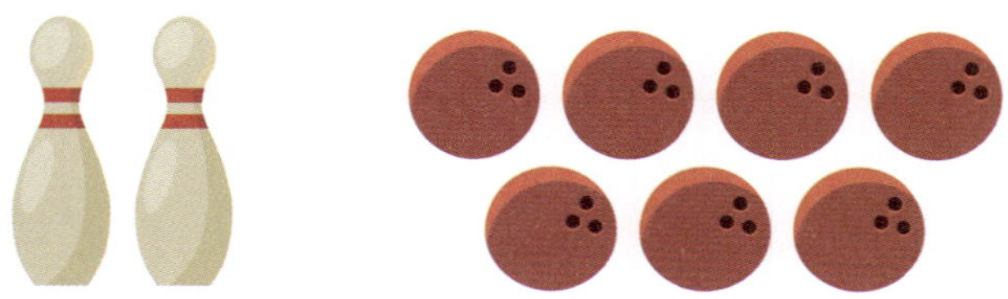

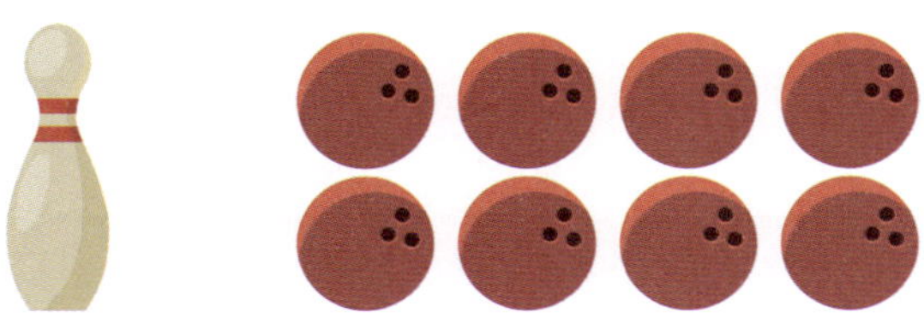

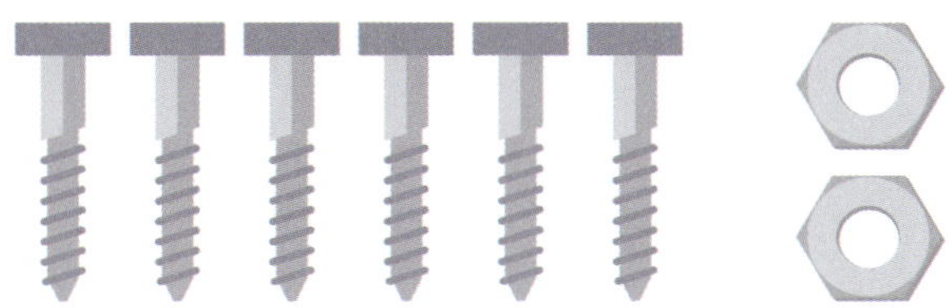

IXL.com skill ID **GBZ**

34 Sums of 8 and 9

Draw to add.

$5 + 4 = 9$

$4 + 4 =$ ____

$2 + 6 =$ ____

Draw to add.

$6 + 3 =$ ____

$7 + 1 =$ ____

$7 + 2 =$ ____

IXL.com
skill ID
WTT

Sums of 8 and 9

Add.

$4 + 4 =$ ______

$3 + 6 =$ ______

$5 + 3 =$ ______

$6 + 2 =$ ______

$2 + 7 =$ ______

$1 + 8 =$ ______

$7 + 1 =$ ______

$5 + 4 =$ ______

$8 + 1 =$ ______

$3 + 5 =$ ______

Let's review! Add.

$1 + 3 =$ ____

$3 + 3 =$ ____

$6 + 3 =$ ____

$2 + 6 =$ ____

$3 + 2 =$ ____

$1 + 7 =$ ____

$7 + 2 =$ ____

$2 + 5 =$ ____

$2 + 4 =$ ____

$4 + 5 =$ ____

IXL.com skill ID TCB

Sums to 9

Add the two numbers on the bottom. Write the total on the top.

8	
3	5

4	2

2	6

3	4

6	1

7	2

Draw a ◯ around the two addition problems that equal the top number.

6

3 + 3	4 + 1
1 + 5	2 + 3

7

4 + 3	2 + 5
4 + 2	3 + 5

5

4 + 2	3 + 1
1 + 4	2 + 3

8

1 + 6	4 + 4
7 + 1	2 + 5

9

5 + 4	6 + 2
3 + 6	4 + 4

IXL.com
skill ID
QJS

Rosa and Lily went to the beach. Write each addition sentence to solve.

Rosa saw 2 sailboats with blue sails. She saw 4 sailboats with white sails. How many sailboats did Rosa see in all?

_______________________________ sailboats

Lily saw 3 crabs by the rocks. She saw 5 more crabs by the dock. How many crabs did Lily see in all?

_______________________________ crabs

Write each addition sentence to solve.

Rosa saw 3 blue fish and 2 yellow fish while she was swimming. How many fish did Rosa see in all?

_______________ fish

Lily saw 6 orange seashells and 2 blue seashells. How many seashells did Lily see in all?

_______________ seashells

Rosa took 4 pictures of starfish. She took 5 pictures of Lily's sand castle. How many pictures did Rosa take in all?

_______________ pictures

IXL.com skill ID **KUH**

Show some ways to make 10. Write each addition sentence.

2 + 8 = 10

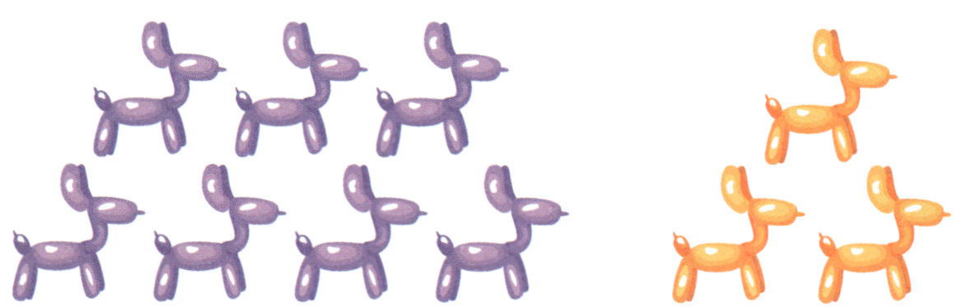

Show some ways to make 10. Write each addition sentence.

IXL.com skill ID CB8

Sums of 10

Find the path from start to finish! Step only on problems that add to 10. No diagonal moves are allowed.

START ↓

9 + 1	2 + 5	4 + 3
2 + 8	5 + 4	6 + 3
4 + 6	1 + 8	2 + 7
5 + 5	3 + 7	3 + 3
6 + 1	1 + 9	8 + 2

FINISH ↓

Let's Learn!

When you add 0, the number stays the same!

3 + 0 = 3

Write each addition sentence.

0 + 2 = 2

Write each addition sentence.

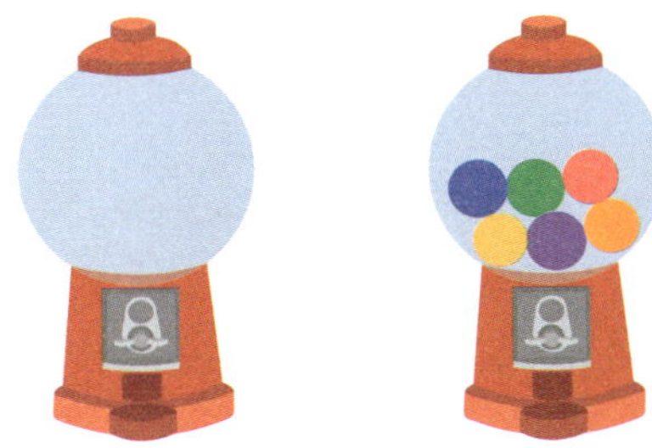

Add.

$1 + 0 =$ ____

$0 + 2 =$ ____

$0 + 4 =$ ____

$9 + 0 =$ ____

$0 + 5 =$ ____

$0 + 7 =$ ____

$0 + 8 =$ ____

$6 + 0 =$ ____

$3 + 0 =$ ____

$10 + 0 =$ ____

IXL.com skill ID **UFN**

Let's review! Add.

$2 + 1 = 3$

$4 + 3 =$ ____

$3 + 2 =$ ____

$0 + 6 =$ ____

$5 + 5 =$ ____

$2 + 7 =$ ____

$8 + 1 =$ ____

$3 + 6 =$ ____

$6 + 2 =$ ____

$3 + 7 =$ ____

Draw lines to add.

5+3

0+2

1+4

4+2

7+3

5

8

2

10

6

Sums to 10

Find the path from start to finish! Add. Follow the right answers to the end.

START ↓

1 + 1	1	2 + 4	6	5 + 1
2		3		6
4 + 4	8	1 + 3	4	5 + 2
7		5		7
7 + 2	8	9 + 1	9	5 + 4
9		10		8
3 + 4	7	FINISH	9	8 + 2

IXL.com skill ID **XWH**

In each group, only two of the addition sentences are right. Cross out the addition sentence that is wrong.

$3 + 2 = 5$

$1 + 2 = 3$

$4 + 1 = 6$

$6 + 2 = 8$

$2 + 2 = 5$

$3 + 3 = 6$

$3 + 6 = 9$

$7 + 1 = 8$

$6 + 4 = 9$

$5 + 5 = 9$

$3 + 5 = 8$

$1 + 8 = 9$

Add. Draw a ◯ around the two problems in each group that have the same answer.

$1 + 4 =$ 5

$2 + 2 =$ 4

$0 + 5 =$ 5

$1 + 6 =$ ____

$5 + 3 =$ ____

$4 + 4 =$ ____

$3 + 4 =$ ____

$5 + 1 =$ ____

$2 + 5 =$ ____

$7 + 2 =$ ____

$5 + 4 =$ ____

$4 + 6 =$ ____

Tic-tac-toe time! Add. Find the row, column, or diagonal where all of the answers are the same.

6 + 2 = 8	7 + 3 = ___	3 + 3 = ___
7 + 1 = ___	5 + 5 = ___	2 + 4 = ___
10 + 0 = ___	2 + 8 = ___	3 + 6 = ___

CHALLENGE ZONE

Add. Draw a ◯ around the larger answer in each row.

$3 + 1 = 4$ $2 + 4 = 6$

$0 + 3 =$ ____ $4 + 1 =$ ____

$1 + 7 =$ ____ $3 + 3 =$ ____

$0 + 9 =$ ____ $2 + 6 =$ ____

Add. Draw a ◯ around the smaller answer in each row.

$1 + 2 = 3$ $1 + 1 = 2$

$7 + 0 =$ ____ $4 + 5 =$ ____

$2 + 3 =$ ____ $3 + 4 =$ ____

$3 + 5 =$ ____ $6 + 1 =$ ____

Missing numbers

Write the missing number.

 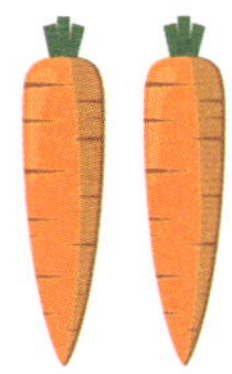

4 + 2 = 6

___ + 1 = 7

2 + ___ = 5

Write the missing number.

____ + 4 = 9

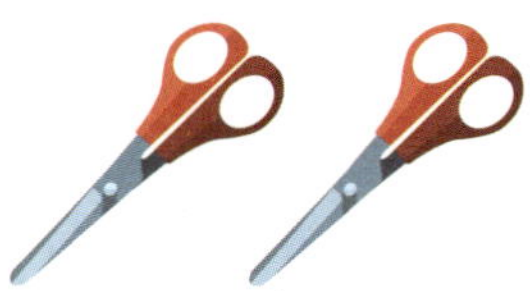 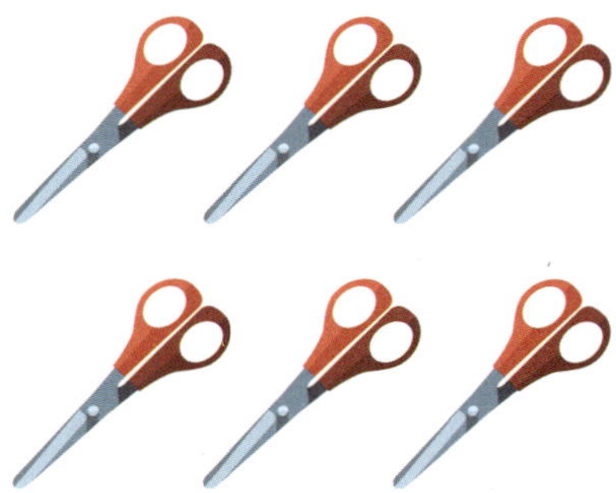

____ + 6 = 8

3 + ____ = 10

58 Missing numbers

Draw objects to find the missing number.

$$2 + 2 = 4$$

$$\_\_ + 4 = 5$$

$$5 + \_\_ = 7$$

Draw objects to find the missing number.

3 + ____ = 7

____ + 4 = 8

Missing numbers

Write the missing number.

$1 + \underline{6} = 7$

$3 + \underline{\quad} = 6$

$\underline{\quad} + 4 = 6$

$\underline{\quad} + 5 = 6$

$6 + \underline{\quad} = 10$

$\underline{\quad} + 3 = 7$

$\underline{\quad} + 0 = 9$

$2 + \underline{\quad} = 5$

$\underline{\quad} + 3 = 10$

$\underline{\quad} + 4 = 8$

Write the missing number.

$0 + \_\_\_ = 3$

$7 + \_\_\_ = 8$

$\_\_\_ + 2 = 7$

$5 + \_\_\_ = 9$

$5 + \_\_\_ = 10$

$6 + \_\_\_ = 6$

$\_\_\_ + 3 = 9$

$\_\_\_ + 7 = 9$

For more practice, visit IXL.com or the IXL mobile app and enter this code in the search bar.

IXL.com skill ID **SAF**

Word problems

CHALLENGE ZONE

Answer each question. Draw a picture to help.

Brandon and Lindsey built a total of 6 snowmen. If Brandon built 4 of the snowmen, how many did Lindsey build?

2 snowmen

Keenan and his sister have a total of 6 star stickers. Keenan has 3 star stickers. How many star stickers does his sister have?

______ star stickers

Krista has 7 bananas. Three of the bananas are on the counter. The rest are in the fridge. How many bananas are in the fridge?

______ bananas

Answer each question. Draw a picture to help.

Erica has 10 balls of yarn. She has 5 balls of yarn that are red. The rest are green. How many of the balls of yarn are green?

_____ balls of yarn

Julie put 7 cupcakes on a plate. There are 2 chocolate cupcakes, and the rest are vanilla. How many cupcakes are vanilla?

_____ cupcakes

Finn has 10 paintbrushes. He dips 2 of them in paint. How many paintbrushes do **not** have paint on them?

_____ paintbrushes

RELATE ADDITION AND SUBTRACTION

Addition and subtraction are related! You can use subtraction to undo addition.

3 + 2 = 5

5 − 2 = 3

TRY IT YOURSELF!

Complete the addition sentence. Then complete the related subtraction sentence.

2 + 4 = ____ 6 − 4 = ____

Complete the addition sentence. Then complete the related subtraction sentence.

$3 + 1 =$ ____ $4 - 1 =$ ____

$6 + 2 =$ ____ $8 - 2 =$ ____

$4 + 3 =$ ____ $7 - 3 =$ ____

$4 + 5 =$ ____ $9 - 5 =$ ____

IXL.com skill ID **DM2**

Adding in any order

Write the addition sentence for each picture.

1 + 3 = 4

3 + 1 = 4

Write the addition sentence for each picture.

TAKE ANOTHER LOOK!

Look at each pair of pictures. The order is swapped. Has the answer changed?

Adding in any order

Add.

$1 + 5 = 6$

$5 + 1 = 6$

$2 + 5 =$ ____

$5 + 2 =$ ____

$3 + 6 =$ ____

$6 + 3 =$ ____

$5 + 0 =$ ____

$0 + 5 =$ ____

$6 + 2 =$ ____

$2 + 6 =$ ____

$3 + 7 =$ ____

$7 + 3 =$ ____

Add. Then write the related addition sentence.

$2 + 0 = 2$

$0 + 2 = 2$

$1 + 7 =$ ____

____

$5 + 4 =$ ____

____

$3 + 5 =$ ____

____

$8 + 2 =$ ____

____

IXL.com skill ID **Y8Y**

Adding three numbers

Let's Learn!

You can add three groups together. Count to find the total!

3 + 1 + 2 = 6

Add.

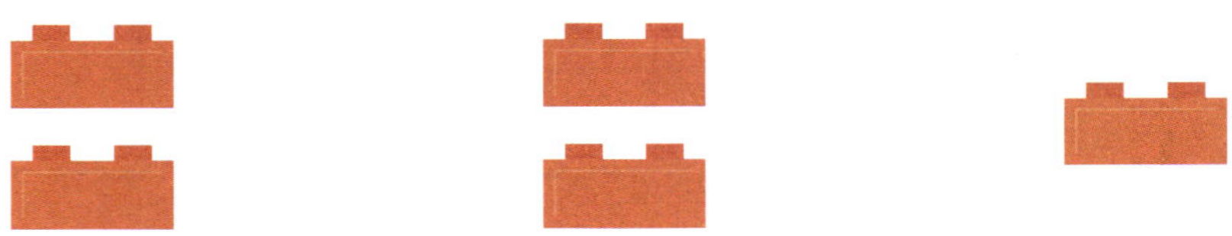

2 + 2 + 1 = 5

___ + ___ + ___ = ___

Add.

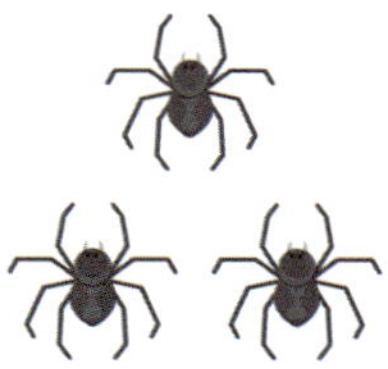  

____ + ____ + ____ = ____

____ + ____ + ____ = ____

____ + ____ + ____ = ____

Draw to add.

$$2 + 1 + 4 = 7$$

$$1 + 3 + 1 =$$

$$2 + 1 + 6 =$$

Draw to add.

$3 + 4 + 1 =$ ______

$3 + 5 + 2 =$ ______

Making teen numbers

Let's Learn!

You can make **teen numbers** with 10 and some more!

10 and 4 is 14

Write the numbers to make each teen number.

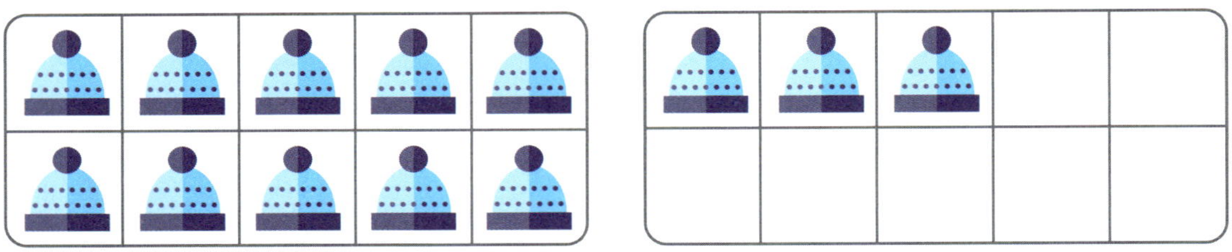

10 and 3 is 13

____ and ____ is ____

Write the numbers to make each teen number.

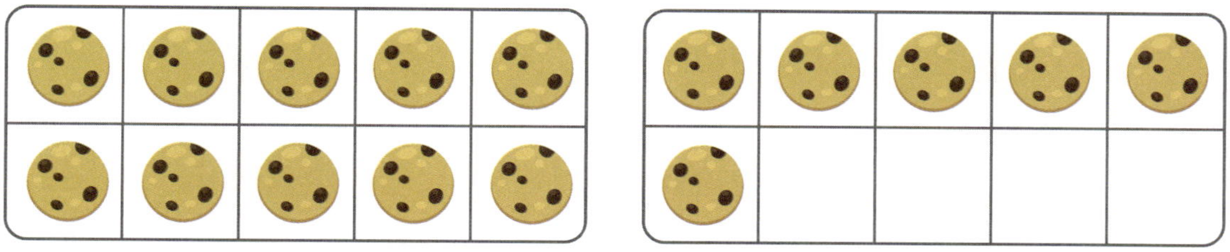

_____ and _____ is _____

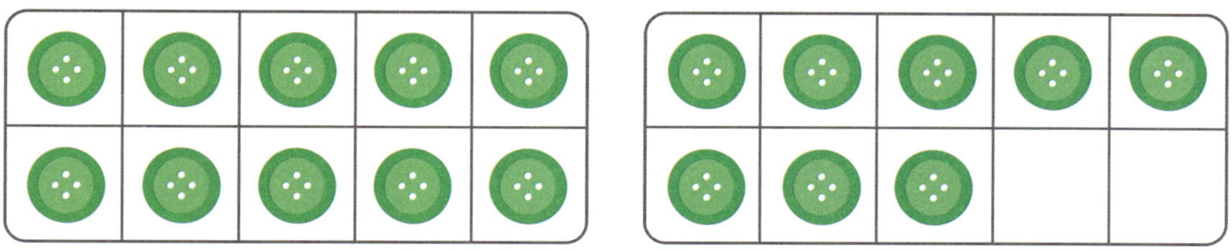

_____ and _____ is _____

_____ and _____ is _____

IXL.com skill ID FTY

Making teen numbers

You can add with 10 to make teen numbers. Fill in the blanks to add.

10 + 2 = 12

____ + ____ = ____

____ + ____ = ____

Fill in the blanks to add.

_____ + _____ = _____

_____ + _____ = _____

_____ + _____ = _____

78 Making teen numbers

Add.

$10 + 2 = 12$

$10 + 4 =$ ______

$10 + 1 =$ ______

$10 + 5 =$ ______

$10 + 8 =$ ______

$10 + 7 =$ ______

$10 + 6 =$ ______

$10 + 9 =$ ______

Write the missing number.

10 + 5 = 15

10 + ____ = 12

____ + 8 = 18

10 + ____ = 17

10 + ____ = 14

____ + 3 = 13

____ + 6 = 16

10 + ____ = 19

80 Adding beyond 10

Write each addition sentence.

$9 + 5 = 14$

Write each addition sentence.

Draw to add.

$9 + 4 = 13$

$6 + 5 =$

$8 + 4 =$

Draw to add.

$3 + 9 =$ ____

$6 + 7 =$ ____

$7 + 4 =$ ____

Emily and her friends are buying food at the fair. Fill in the blanks to solve.

Emily wants to buy popcorn and a pretzel. How much money will she need?

$ 2 + $ 3 = $ 5

Carmen wants to buy a pretzel and a candy apple. How much money will she need?

$ ____ + $ ____ = $ ____

Fill in the blanks to solve.

Henry wants to buy a candy apple and popcorn. How much money will he need?

\$ ______ + \$ ______ = \$ ______

Kimi wants to buy ice cream and a corn dog. How much money will she need?

\$ ______ + \$ ______ = \$ ______

Owen wants to buy 2 corn dogs. How much money will he need?

\$ ______ + \$ ______ = \$ ______

Cooper and his friends are at Toys Ahoy. Fill in the blanks to solve.

Cooper wants to buy a teddy bear and a xylophone. How much money will he need?

\$ ______ + \$ ______ = \$ ______

Zoe wants to buy 2 xylophones. How much money will she need?

\$ ______ + \$ ______ = \$ ______

Answer each question. Draw a ◯ around *yes* or *no*.

Paco wants to buy 2 teddy bears. He has $5. Does he have enough money?

yes no

Abby wants to buy a doll and a xylophone. She has $8. Does she have enough money?

yes no

Liz wants to buy a xylophone and a helicopter. She has $10. Does she have enough money?

yes no

Brody wants to buy a helicopter for himself. He also wants to buy a teddy bear for his sister. He has $8. Does he have enough money?

yes no

Review your addition facts. Try to be quick! Add.

$2 + 1 =$ ______

$1 + 1 =$ ______

$1 + 3 =$ ______

$2 + 3 =$ ______

$5 + 0 =$ ______

$2 + 2 =$ ______

$7 + 1 =$ ______

$1 + 5 =$ ______

$4 + 3 =$ ______

$2 + 4 =$ ______

$2 + 5 =$ ______

$5 + 3 =$ ______

Keep going! Add.

3 + 3 = ____

0 + 9 = ____

4 + 5 = ____

1 + 6 = ____

9 + 1 = ____

4 + 4 = ____

3 + 6 = ____

6 + 4 = ____

5 + 5 = ____

2 + 6 = ____

For more practice, visit IXL.com or the IXL mobile app and enter this code in the search bar.

IXL.com skill ID **TCB**

Sums to 10

Draw lines to add.

5 + 4	5
1 + 4	10
5 + 2	6
7 + 3	9
3 + 5	7
4 + 2	8

Find the path from start to finish! Add. Follow the right answers to the end.

START ↓

1 + 6	7	1 + 0	0	4 + 4
6		1		7
1 + 9	5	3 + 2	6	5 + 5
10		4		9
7 + 2	8	3 + 3	7	4 + 6
9		5		9
2 + 8	10	FINISH	6	3 + 4

92 Word problems

Write each addition sentence to solve.

Mr. Davis planted 2 lemon trees and 5 orange trees. How many trees did he plant in all?

__________________________________ trees

Ann cleaned her room. She found 1 building block on her dresser. She found 4 more building blocks under her bed. How many building blocks did Ann find?

__________________________________ building blocks

Mitch and Steve are playing checkers. Mitch wins 4 games. Steve wins 5 games. How many games do they play in all?

__________________________________ games

Write each addition sentence to solve.

Dalton is making a fort in his room. He uses 4 blue blankets and 3 green blankets. How many blankets does Dalton use?

______________________ blankets

Belle is at the park. She goes down the red slide 6 times. She goes down the yellow slide 2 times. How many times is that in all?

______________________ times

Miguel saw a large boat and a small boat out on the lake. There were 6 people on the large boat and 3 people on the small boat. How many people is that in all?

______________________ people

IXL.com skill ID **ZE8**

Missing numbers

Draw spots to find the missing number.

$3 + 2 = 5$

$4 + \_\_\_ = 7$

$2 + \_\_\_ = 8$

$6 + \_\_\_ = 10$

Write the missing number.

$2 + \_\_\_ = 6$

$\_\_\_ + 3 = 6$

$\_\_\_ + 5 = 8$

$\_\_\_ + 4 = 5$

$5 + \_\_\_ = 10$

$\_\_\_ + 7 = 9$

$6 + \_\_\_ = 6$

$4 + \_\_\_ = 10$

$8 + \_\_\_ = 10$

CHALLENGE ZONE

Use the numbers to fill in the blanks. Use each number one time.

2, 3, 6, 1

3 + 1 = 4

2 + 6 = 8

1, 6, 2, 3

___ + ___ = 5

___ + ___ = 7

Use the numbers to fill in the blanks. Use each number one time.

7, 4, 6, 2

_____ + _____ = 9

_____ + _____ = 10

2, 3, 4, 5

_____ + _____ = 6

_____ + _____ = 8

98 Making teen numbers

Add.

$10 + 5 =$ ____

$10 + 4 =$ ____

$10 + 2 =$ ____

$10 + 1 =$ ____

$10 + 3 =$ ____

$10 + 7 =$ ____

$10 + 8 =$ ____

$10 + 6 =$ ____

Write the missing number.

10 + ____ = 11

10 + ____ = 13

____ + 2 = 12

10 + ____ = 15

10 + 4 = ____

10 + 8 = ____

____ + 7 = 17

10 + ____ = 19

10 + 6 = ____

____ + 1 = 11

Brett's Beach Shop sells items for the beach. Fill in the blanks to solve.

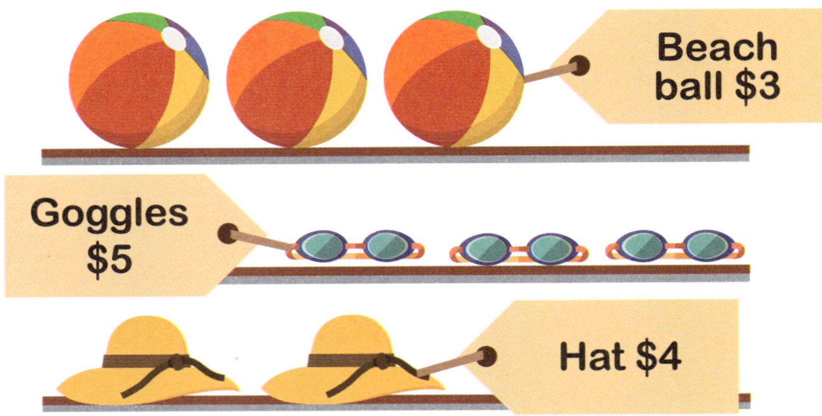

Chris wants to buy goggles and a beach ball. How much money will he need?

\$ ______ + \$ ______ = \$ ______

Ellie wants to buy 2 beach balls. How much money will she need?

\$ ______ + \$ ______ = \$ ______

Mrs. Williams wants to buy 2 hats for her friends. How much money will she need?

\$ ______ + \$ ______ = \$ ______

Ethan, Nina, and Jen save money in piggy banks. Look at the picture to see how much they have saved. Fill in the blanks to solve.

Nina's dad gives her $3. How much money does she have now?

\$ ______ + \$ ______ = \$ ______

Ethan adds $4 to his piggy bank. How much money does he have now?

\$ ______ + \$ ______ = \$ ______

Jen earns $8 selling lemonade. How much money does she have now?

\$ ______ + \$ ______ = \$ ______

Lucy and Ruben are making their own fruit salads. Answer each question.

Lucy's Fruit Salad

2 bananas

4 oranges

3 apples

8 strawberries

Ruben's Fruit Salad

4 bananas

3 oranges

6 apples

1 pineapple

How many bananas will Lucy and Ruben need in all?

__________ bananas

How many apples will Lucy and Ruben need in all?

__________ apples

Answer each question.

Lucy peels 2 oranges. How many more oranges does she need to peel for her fruit salad?

_____________ oranges

Ruben has 2 apples right now. How many more does he need for his fruit salad?

_____________ apples

Lucy has a big basket of strawberries. She needs to put 1 more into the fruit salad. How many strawberries are already in the salad?

_____________ strawberries

Answer key

PAGE 2

4

5

2

6

10

PAGE 3

3

1

4

9

7

8

PAGE 4

1 and 3 is 4

2 and 3 is 5

2 and 5 is 7

PAGE 5

5 and 1 is 6

4 and 4 is 8

4 and 5 is 9

PAGE 6

2 and 4 is 6

4 and 1 is 5

3 and 3 is 6

PAGE 7

3 and 5 is 8

6 and 3 is 9

PAGE 8

3 and 4 is 7

7 and 2 is 9

6 and 4 is 10

PAGE 9

6 and 2 is 8

5 and 5 is 10

PAGE 10

2 + 1 is 3

4 + 2 is 6

PAGE 11

2 + 2 is 4

3 + 4 is 7

6 + 1 is 7

PAGE 12

1 + 4 = 5

6 + 2 = 8

PAGE 13

3 + 3 = 6

5 + 3 = 8

4 + 5 = 9

PAGE 14

3 + 1 = 4 dogs

4 + 2 = 6 flowers

PAGE 15

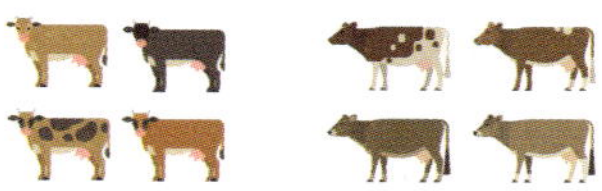

4 + 4 = 8 cows

2 + 6 = 8 fish

PAGE 16

3+2=5

2+2=4

1+3=4

2+3=5

4+1=5

3+1=4

PAGE 17

4 + 1 = 5

2 + 2 = 4

3 + 1 = 4

2 + 3 = 5

PAGE 18

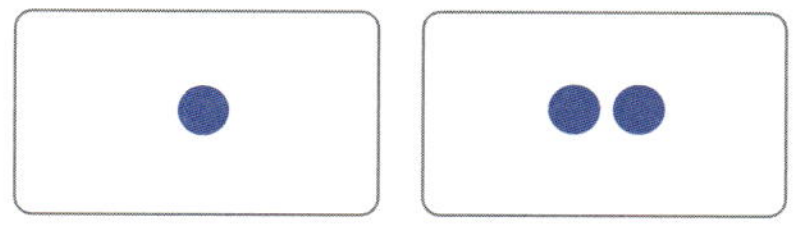

1 + 2 = 3

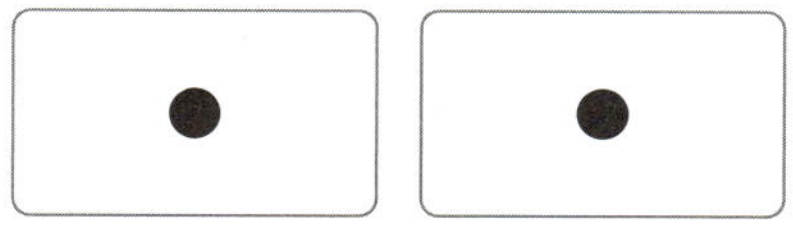

1 + 1 = 2

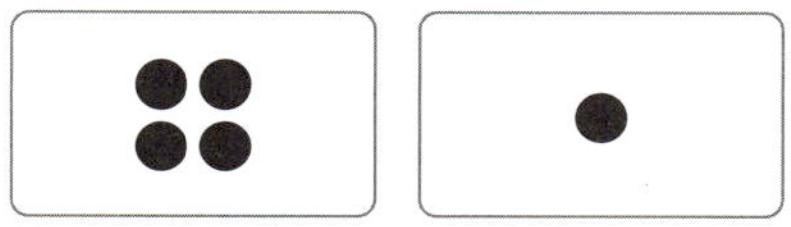

4 + 1 = 5

PAGE 19

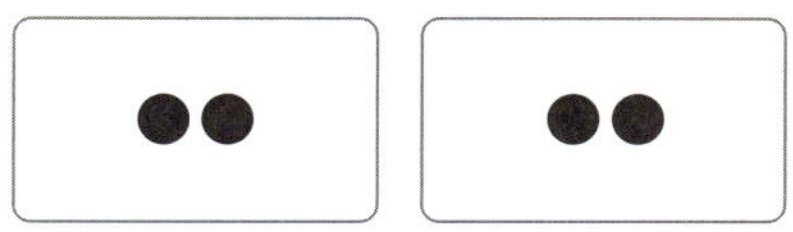

2 + 2 = 4

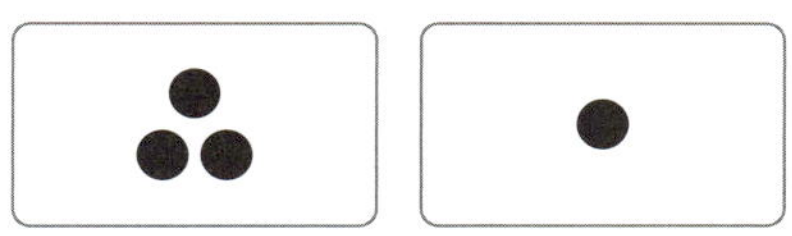

3 + 1 = 4

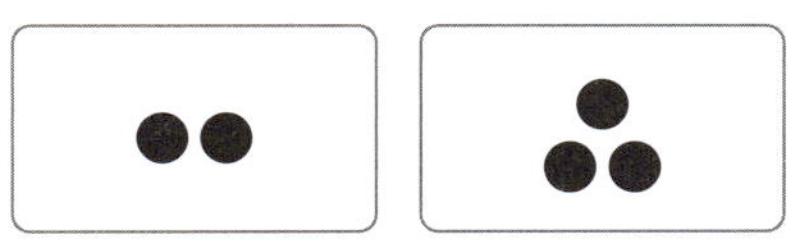

2 + 3 = 5

PAGE 20

1 + 4 = 5

1 + 1 = 2

1 + 2 = 3

3 + 1 = 4

2 + 2 = 4

2 + 3 = 5

PAGE 21

1 + 3 = 4

4 + 1 = 5

2 + 1 = 3

1 + 1 = 2

3 + 2 = 5

2 + 2 = 4

1 + 2 = 3

2 + 3 = 5

1 + 4 = 5

3 + 1 = 4

PAGE 22

1 + 2 = 3 eggs

2 + 2 = 4 candles

PAGE 23

3 + 1 = 4 handprints

2 + 3 = 5 kites

PAGE 24

3+4=7

3+3=6

5+1=6

1+6=7

2+5=7

2+4=6

Answer key

PAGE 25

4+2=6

4+3=7

6+1=7

1+5=6

PAGE 26

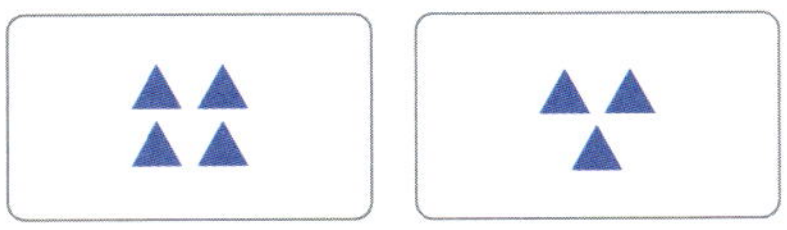

4 + 3 = 7

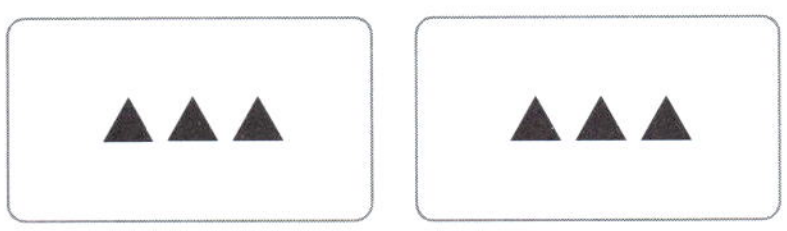

3 + 3 = 6

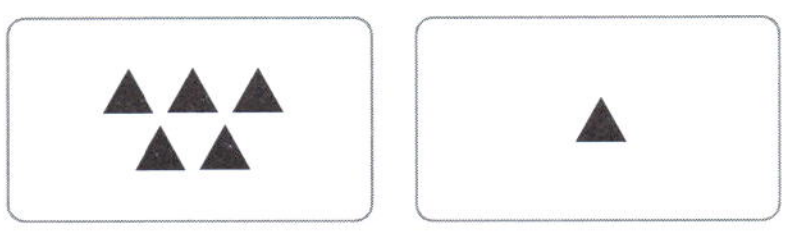

5 + 1 = 6

PAGE 27

5 + 2 = 7

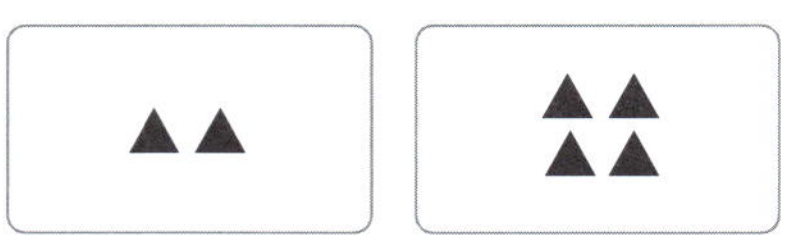

2 + 4 = 6

6 + 1 = 7

PAGE 28

2 + 4 = 6

4 + 3 = 7

5 + 1 = 6

5 + 2 = 7

6 + 1 = 7

3 + 3 = 6

3 + 4 = 7

4 + 2 = 6

1 + 5 = 6

2 + 5 = 7

PAGE 29

1 + 2 = 3

3 + 1 = 4

3 + 4 = 7

1 + 5 = 6

2 + 2 = 4

4 + 1 = 5

3 + 2 = 5

2 + 5 = 7

1 + 6 = 7

4 + 2 = 6

PAGE 30

1 + 3 = 4 cakes

4 + 2 = 6 turtles

5 + 2 = 7 pine cones

PAGE 31

3 + 2 = 5 eggs

4 + 3 = 7 apples

PAGE 32

4+4=8

5+4=9

8+1=9

2+6=8

3+5=8

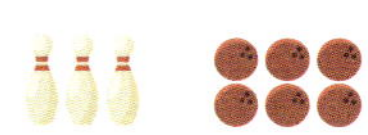

3+6=9

PAGE 33

4+5=9

1+7=8

5+3=8

2+7=9

1+8=9

6+2=8

PAGE 34

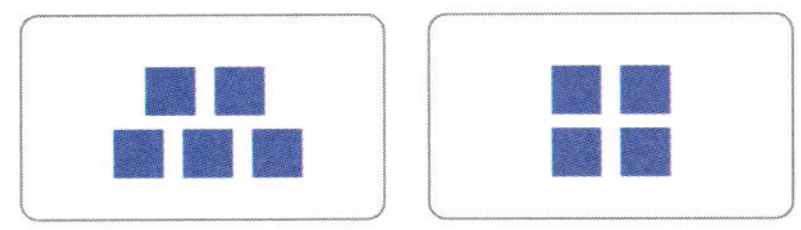

5 + 4 = 9

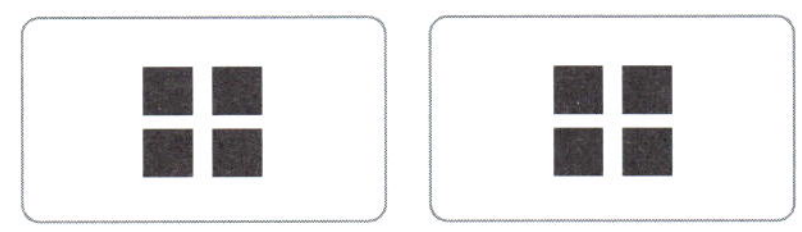

4 + 4 = 8

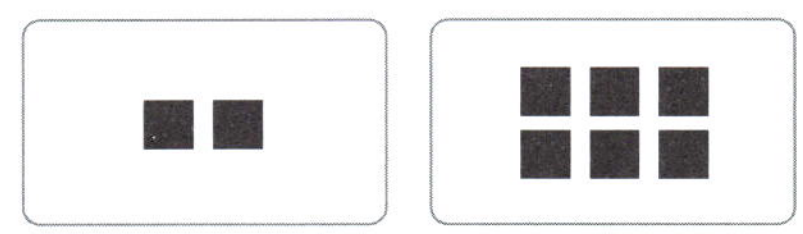

2 + 6 = 8

PAGE 35

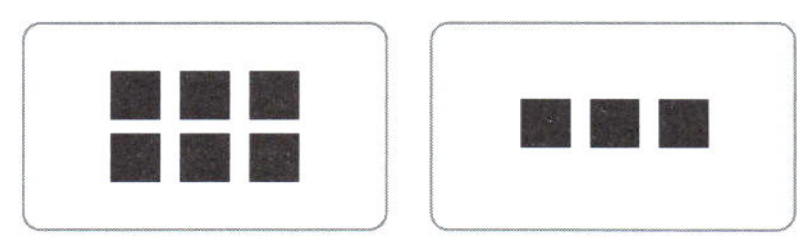

6 + 3 = 9

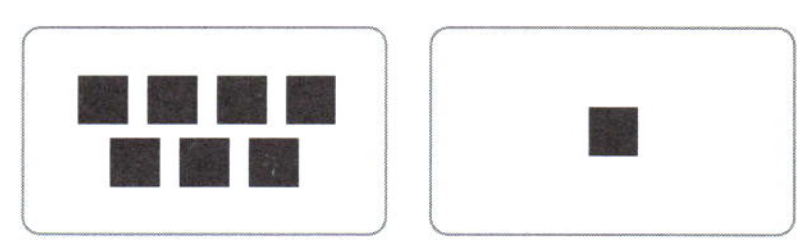

7 + 1 = 8

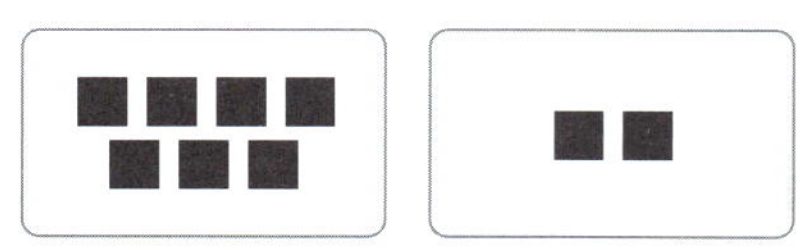

7 + 2 = 9

PAGE 36

4 + 4 = 8
5 + 3 = 8
2 + 7 = 9
7 + 1 = 8
8 + 1 = 9

3 + 6 = 9
6 + 2 = 8
1 + 8 = 9
5 + 4 = 9
3 + 5 = 8

PAGE 37

1 + 3 = 4
6 + 3 = 9
3 + 2 = 5
7 + 2 = 9
2 + 4 = 6

3 + 3 = 6
2 + 6 = 8
1 + 7 = 8
2 + 5 = 7
4 + 5 = 9

PAGE 38

8	
3	5

6	
4	2

8	
2	6

7	
3	4

7	
6	1

9	
7	2

PAGE 39

6
3 + 3 4 + 1
1 + 5 2 + 3

7
4 + 3 2 + 5
4 + 2 3 + 5

5
4 + 2 3 + 1
1 + 4 2 + 3

8
1 + 6 4 + 4
7 + 1 2 + 5

9
5 + 4 6 + 2
3 + 6 4 + 4

PAGE 40

2 + 4 = 6 sailboats

3 + 5 = 8 crabs

PAGE 41

3 + 2 = 5 fish

6 + 2 = 8 seashells

4 + 5 = 9 pictures

PAGE 42

2 + 8 = 10

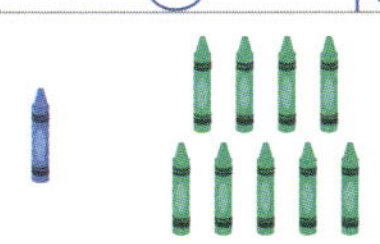

1 + 9 = 10

7 + 3 = 10

PAGE 43

5 + 5 = 10

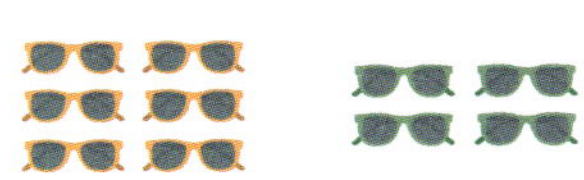

6 + 4 = 10

8 + 2 = 10

PAGE 44

START

9 + 1	2 + 5	4 + 3
2 + 8	5 + 4	6 + 3
4 + 6	1 + 8	2 + 7
5 + 5	3 + 7	3 + 3
6 + 1	1 + 9	8 + 2

FINISH

PAGE 45

0 + 2 = 2 5 + 0 = 5

0 + 4 = 4 7 + 0 = 7

PAGE 46

0 + 3 = 3 4 + 0 = 4

2 + 0 = 2 0 + 6 = 6

PAGE 47

1 + 0 = 1
0 + 4 = 4
0 + 5 = 5
0 + 8 = 8
3 + 0 = 3

0 + 2 = 2
9 + 0 = 9
0 + 7 = 7
6 + 0 = 6
10 + 0 = 10

Answer key

PAGE 48

2 + 1 = 3
3 + 2 = 5
5 + 5 = 10
8 + 1 = 9
6 + 2 = 8

4 + 3 = 7
0 + 6 = 6
2 + 7 = 9
3 + 6 = 9
3 + 7 = 10

PAGE 49

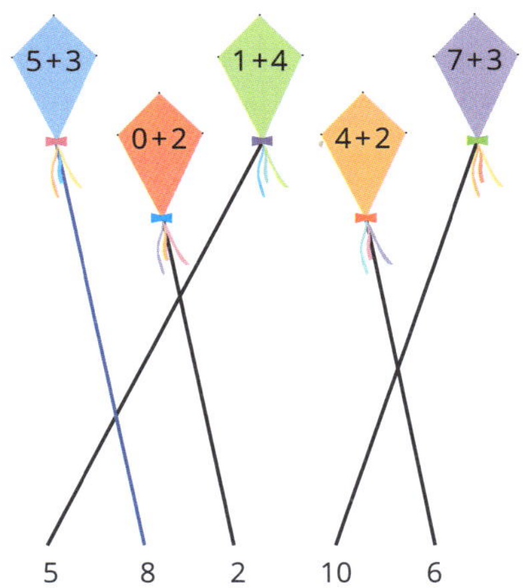

PAGE 50

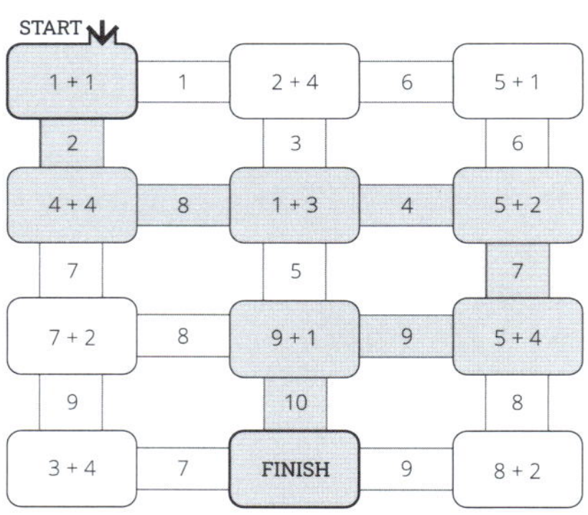

PAGE 51

3 + 2 = 5
1 + 2 = 3
~~4 + 1 = 6~~

6 + 2 = 8
~~2 + 2 = 5~~
3 + 3 = 6

3 + 6 = 9
7 + 1 = 8
~~6 + 4 = 9~~

~~5 + 5 = 9~~
3 + 5 = 8
1 + 8 = 9

PAGE 52

1 + 4 = 5
2 + 2 = 4
0 + 5 = 5

1 + 6 = 7
5 + 3 = 8
4 + 4 = 8

3 + 4 = 7
5 + 1 = 6
2 + 5 = 7

7 + 2 = 9
5 + 4 = 9
4 + 6 = 10

PAGE 53

6 + 2 = 8	7 + 3 = 10	3 + 3 = 6
7 + 1 = 8	5 + 5 = 10	2 + 4 = 6
10 + 0 = 10	2 + 8 = 10	3 + 6 = 9

PAGE 54

3 + 1 = 4
0 + 3 = 3
1 + 7 = 8
0 + 9 = 9

2 + 4 = 6
4 + 1 = 5
3 + 3 = 6
2 + 6 = 8

PAGE 55

1 + 2 = 3
7 + 0 = 7
2 + 3 = 5
3 + 5 = 8

1 + 1 = 2
4 + 5 = 9
3 + 4 = 7
6 + 1 = 7

PAGE 56

4 + 2 = 6

6 + 1 = 7

2 + 3 = 5

PAGE 57

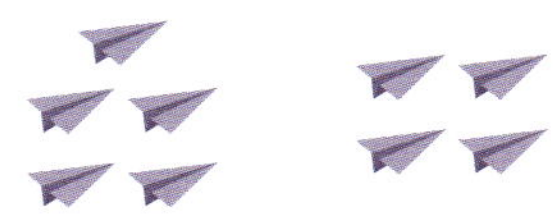

5 + 4 = 9

2 + 6 = 8

3 + 7 = 10

PAGE 58

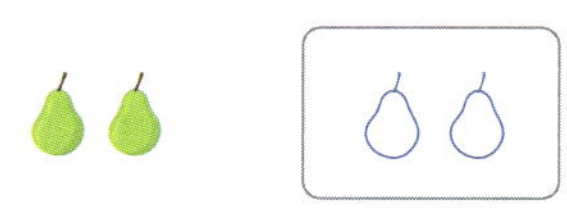

2 + 2 = 4

1 + 4 = 5

5 + 2 = 7

PAGE 59

3 + 4 = 7

4 + 4 = 8

PAGE 60

1 + 6 = 7　　3 + 3 = 6

2 + 4 = 6　　1 + 5 = 6

6 + 4 = 10　　4 + 3 = 7

9 + 0 = 9　　2 + 3 = 5

7 + 3 = 10　　4 + 4 = 8

PAGE 61

0 + 3 = 3　　7 + 1 = 8

5 + 2 = 7　　5 + 4 = 9

5 + 5 = 10　　6 + 0 = 6

6 + 3 = 9　　2 + 7 = 9

PAGE 62

2 snowmen

3 star stickers

4 bananas

PAGE 63

5 balls of yarn

5 cupcakes

8 paintbrushes

PAGE 64

2 + 4 = 6　　6 − 4 = 2

PAGE 65

3 + 1 = 4　　4 − 1 = 3

6 + 2 = 8　　8 − 2 = 6

4 + 3 = 7　　7 − 3 = 4

4 + 5 = 9　　9 − 5 = 4

PAGE 66

1+3=4　　3+1=4

2+1=3　　1+2=3

2+3=5　　3+2=5

PAGE 67

2+4=6　　4+2=6

1+4=5　　4+1=5

5+1=6　　1+5=6

No. The answer is the same.

PAGE 68

1 + 5 = 6　　2 + 5 = 7

5 + 1 = 6　　5 + 2 = 7

3 + 6 = 9　　5 + 0 = 5

6 + 3 = 9　　0 + 5 = 5

6 + 2 = 8　　3 + 7 = 10

2 + 6 = 8　　7 + 3 = 10

PAGE 69

2 + 0 = 2　　1 + 7 = 8

0+2=2　　7+1=8

5 + 4 = 9　　3 + 5 = 8

4+5=9　　5+3=8

8 + 2 = 10

2+8=10

PAGE 70

2 + 2 + 1 = 5

1 + 2 + 1 = 4

PAGE 71

3 + 1 + 3 = 7

4 + 2 + 2 = 8

5 + 1 + 3 = 9

PAGE 72

2 + 1 + 4 = 7

1 + 3 + 1 = 5

2 + 1 + 6 = 9

Answer key

PAGE 73

3 + 4 + 1 = 8

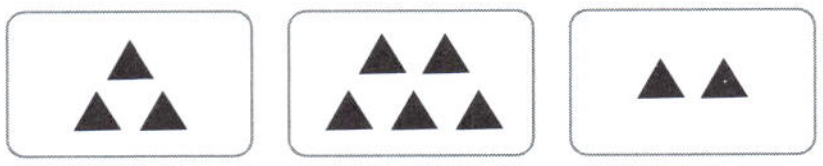

3 + 5 + 2 = 10

PAGE 74

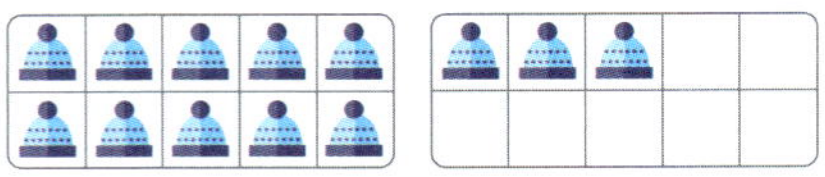

10 and 3 is 13

10 and 5 is 15

PAGE 75

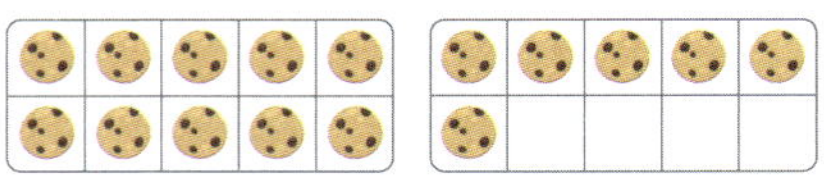

10 and 6 is 16

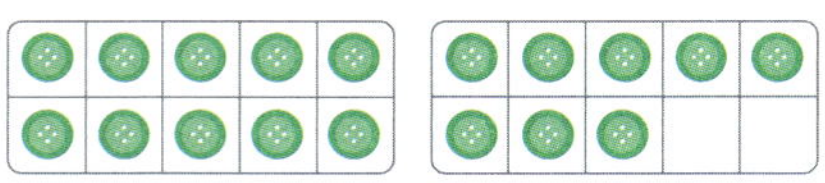

10 and 8 is 18

10 and 7 is 17

PAGE 76

10 + 2 = 12

10 + 6 = 16

PAGE 76, *continued*

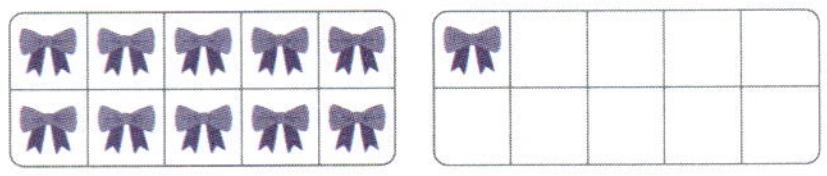

10 + 1 = 11

PAGE 77

10 + 5 = 15

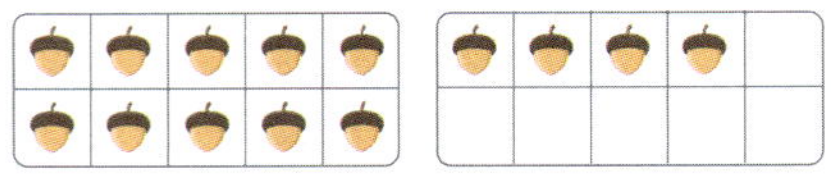

10 + 4 = 14

10 + 9 = 19

PAGE 78

10 + 2 = 12

10 + 4 = 14

10 + 1 = 11

10 + 5 = 15

10 + 8 = 18

10 + 7 = 17

10 + 6 = 16

10 + 9 = 19

PAGE 79

10 + 5 = 15

10 + 2 = 12

10 + 8 = 18

10 + 7 = 17

10 + 4 = 14

10 + 3 = 13

10 + 6 = 16

10 + 9 = 19

PAGE 80

9 + 5 = 14

8 + 3 = 11

6 + 7 = 13

PAGE 81

4 + 8 = 12

6 + 9 = 15

2 + 9 = 11

PAGE 82

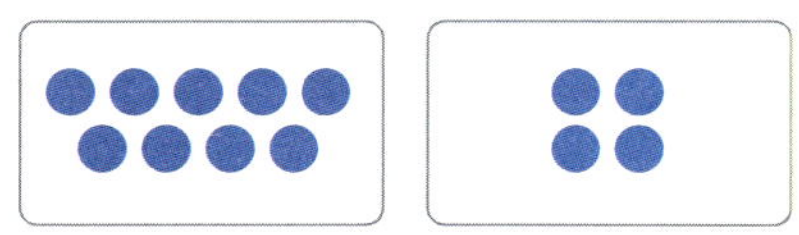

9 + 4 = 13

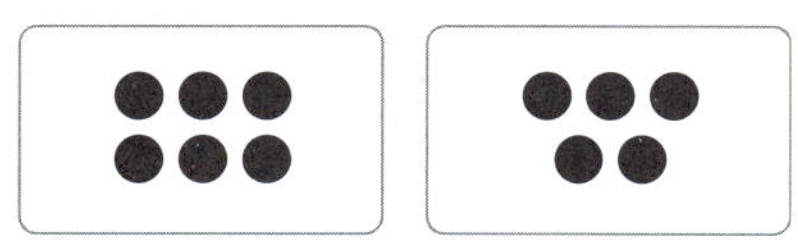

6 + 5 = 11

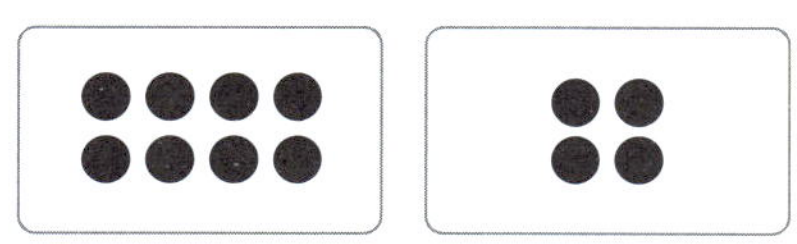

8 + 4 = 12

Page 83

3 + 9 = 12

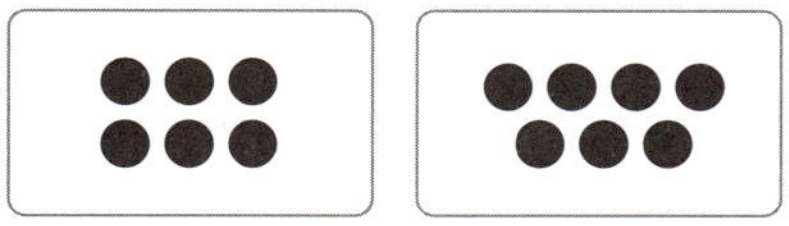

6 + 7 = 13

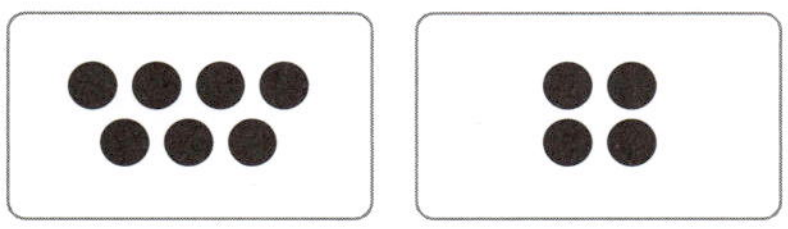

7 + 4 = 11

Page 84

\$ 2 + \$ 3 = \$ 5

\$ 3 + \$ 6 = \$ 9

PAGE 85

\$ 6 + \$ 2 = \$ 8

\$ 4 + \$ 5 = \$ 9

\$ 5 + \$ 5 = \$ 10

PAGE 86

\$ 2 + \$ 3 = \$ 5

\$ 3 + \$ 3 = \$ 6

PAGE 87

yes

no

yes

no

PAGE 88

2 + 1 = 3

1 + 3 = 4

5 + 0 = 5

7 + 1 = 8

4 + 3 = 7

2 + 5 = 7

1 + 1 = 2

2 + 3 = 5

2 + 2 = 4

1 + 5 = 6

2 + 4 = 6

5 + 3 = 8

PAGE 89

3 + 3 = 6

4 + 5 = 9

9 + 1 = 10

3 + 6 = 9

5 + 5 = 10

0 + 9 = 9

1 + 6 = 7

4 + 4 = 8

6 + 4 = 10

2 + 6 = 8

PAGE 90

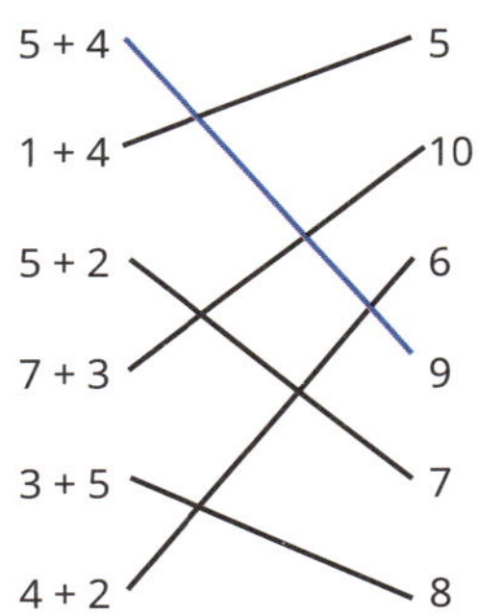

PAGE 91

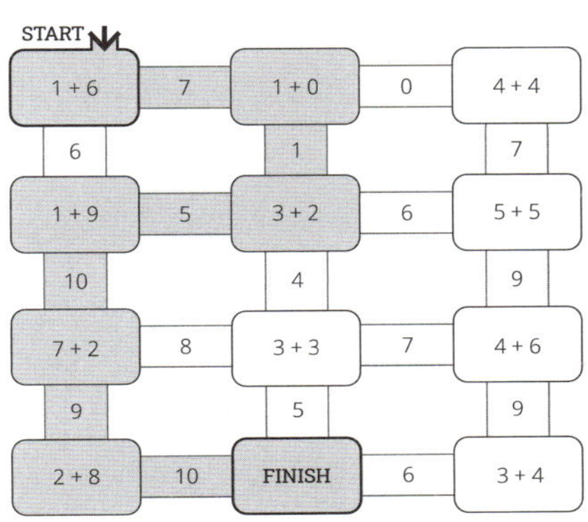

PAGE 92

2 + 5 = 7 trees

1 + 4 = 5 building blocks

4 + 5 = 9 games

PAGE 93

4 + 3 = 7 blankets

6 + 2 = 8 times

6 + 3 = 9 people

PAGE 94

3 + 2 = 5

4 + 3 = 7

2 + 6 = 8

6 + 4 = 10

PAGE 95

2 + 4 = 6

3 + 5 = 8

5 + 5 = 10

6 + 0 = 6

8 + 2 = 10

3 + 3 = 6

1 + 4 = 5

2 + 7 = 9

4 + 6 = 10

PAGE 96

Order of addends may vary.

2, 3, 6, 1

3 + 1 = 4

2 + 6 = 8

1, 6, 2, 3

2 + 3 = 5

1 + 6 = 7

Answer key

PAGE 97
Order of addends may vary.

7, 4, 6, 2

7 + 2 = 9
6 + 4 = 10

2, 3, 4, 5

2 + 4 = 6
3 + 5 = 8

PAGE 98

10 + 5 = 15
10 + 4 = 14
10 + 2 = 12
10 + 1 = 11
10 + 3 = 13
10 + 7 = 17
10 + 8 = 18
10 + 6 = 16

PAGE 99

10 + 1 = 11
10 + 3 = 13
10 + 2 = 12
10 + 5 = 15
10 + 4 = 14
10 + 8 = 18
10 + 7 = 17
10 + 9 = 19
10 + 6 = 16
10 + 1 = 11

PAGE 100

\$5 + \$3 = \$8
\$3 + \$3 = \$6
\$4 + \$4 = \$8

PAGE 101

\$7 + \$3 = \$10
\$3 + \$4 = \$7
\$2 + \$8 = \$10

PAGE 102

6 bananas
9 apples

PAGE 103

2 oranges
4 apples
7 strawberries